ENERGY SCIENCE, ENGINEERING AND TECHNOLOGY

UTILISATION AND DEVELOPMENT OF SOLAR AND WIND RESOURCES

ENERGY SCIENCE, ENGINEERING AND TECHNOLOGY

Additional books in this series can be found on Nova's website at:

https://www.novapublishers.com/catalog/index.php?cPath=23_29&seriesp=Energy+Science%2C+Engineering+and+Technology

Additional E-books in this series can be found on Nova's website at:

https://www.novapublishers.com/catalog/index.php?cPath=23_29&seriespe=Energy+Science%2C+Engineering+and+Technology

ENERGY SCIENCE, ENGINEERING AND TECHNOLOGY

UTILISATION AND DEVELOPMENT OF SOLAR AND WIND RESOURCES

Abdeen Mustafa Omer

Nova Science Publishers, Inc.
New York

For permission to use material from this book please contact us:
Telephone 631-231-7269; Fax 631-231-8175
Web Site: http://www.novapublishers.com

LIBRARY OF CONGRESS CATALOGING-IN-PUBLICATION DATA

Omer, Abdeen Mustafa.
Utilisation and development of solar and wind resources / editor, Abdeen Mustafa Omer.
p. cm. -- (Energy science, engineering and technology)
Includes bibliographical references and index.
ISBN 978-1-61668-238-5 (pbk.)
1. Renewable energy sources. 2. Solar power. 3. Wind power. 4. Energy development. I. Title.
TJ808.O48 2011
333.79'4--dc23

2011039222

Published by Nova Science Publishers, Inc. † New York

CONTENTS

ABSTRACT

The imminent exhaustion of fossil energy resources and the increasing demand for energy were the motives for those reasonable in world to put into practice an energy policy based on rational use of energy; and on exploitation of new and renewable energy sources. After 1980, as the supply of conventional energy has not been able to follow the tremendous increase of the production demand in rural areas of the world, a renewed interest for the application of solar and wind energy has shown in many places. Therefore, the researchers and engineers began to pay more attention to wind and solar energy utilisation in rural areas. Because the wind energy resource in many rural areas is sufficient for attractive application of wind pumps, and as fuel is insufficient, the wind pumps will be spread on a rather large-scale in the near future. Wind is a form of renewable energy, which is always in a non-steady state due to the wide temporal and spatial variations of wind velocity. The need for the provision of new data stations in order to enable a complete and reliable assessment of the overall wind power potential of the country is to be identified and specific locations suggested. This book presents the background and ideas of the development of the concept as well as the main results and experience gained during ongoing projects up to now. In the world, various designs of wind machines for water pumping have been developed and some designs are presently manufactured commercially. Results suggest that wind power would be more profitably used for local and small-scale applications especially for remote rural areas. It is concluded that many parts of the world is enjoyed with abundant wind and solar energy resources.

Keywords: Renewable energy technologies, wind and solar resources, utilisation and development.

Chapter 1

INTRODUCTION

New and renewable sources of energy can make an increasing contribution to the energy supply mix of the world in view of favourable renewable energy resource endowments, limitations and uncertainties of fossil fuel supplies, adverse balance of payments and the increasing pressure on environment from conventional energy generation. Among the renewable energy technologies, the generation of mechanical and electrical power by wind machines has emerged as a techno-economical viable and cost-effective option. The provision of pumped clean water is one of the best ways to improve health and increase the productive capacity of the population. Rural access to clean water is best achieved through pumping from underground water aquifers rather than using surface water sources, which are often polluted. Because of the relatively small quantities of water required, wind pumping for village water supply and livestock watering can, is cost-effective given a good wind site. Irrigation pumping however requires large quantities of water at specific times of the year. For much of the year the pump may be idle or oversized and wind pumping for irrigation may be more difficult to justify on economic grounds. Wind energy is one of the several energy sources, alternatives to the conventional primary energy resources, which are now power man's industrial and socio-economic activities worldwide. With the notable exception of hydropower (which in fact is renewable), these primary energy resources have definite lifetimes, and depending on use rate among several other factors.

The sources to alleviate the energy situation in the world are sufficient to supply all foreseeable needs. Conservation of energy and rationing in some form will however have to be practised by most countries, to reduce oil imports and redress balance of payments positions. Meanwhile development and application of nuclear power and some of the traditional solar, wind and water energy alternatives must be set in hand to supplement what remains of the fossil fuels. The encouragement of greater energy use is an essential component of development. In the short-term, it requires mechanisms to enable the rapid increase in energy/capita, and in the long-term, we should be working towards a way of life, which makes use of energy efficiency and without the impairment of the environment or of causing safety problems. Such a programme should as far as possible be based on renewable energy resources.

Large-scale, conventional, power plant such as hydropower, has an important part to play in development. It does not, however, provide a complete solution. There is an important complementary role for the greater use of small-scale, rural based and power plants. Such plant can be used to assist development since it can be made locally using local resources, enabling a rapid built-up in total equipment to be made without a corresponding and unacceptably large demand on central funds. Renewable resources are particularly suitable for providing the energy for such equipment and its use is also compatible with the long-term aims. It is possible with relatively simple flat plate solar collectors to provide warmed water and enable some space heating for homes and offices which is particularly useful when the buildings are well insulated and thermal capacity sufficient for the carry over of energy from day to night is arranged. In addition to the drain on resources, such an increase in consumption consequences, together with the increased hazards of pollution and the safety problems associated with a large nuclear fission programmes are the problems we face. This is a disturbing prospect. It would be equally unacceptable to suggest that the difference in energy between the developed and developing countries and prudent for the developed countries to move towards a way of life which, whilst maintaining or even increasing quality of life, and reduce significantly the energy consumption per capita. Such savings can be achieved in a number of ways:

Improved efficiency of energy use, for example better thermal insulation, energy recovery, and total energy.

Conservation of energy resources by design for long life and recycling rather than the short life throwaway product.

Systematic replanning of our way of life, for example in the field of transport. A strong local political commitment to the environment and sustainability.

Currently the non-commercial fuels wood, crop residues and animal dung are used in large amounts in the rural areas of developing countries, principally for heating and cooking, the method of use is highly inefficient. As in the developed countries, the fossil fuels are currently of great importance in the developing countries. Geothermal and tidal energy are less important though, of course, will have local significance where conditions are suitable. Nuclear energy sources are included for completeness, but are not likely to make any effective contribution in the rural areas.

Between 1980 and 2000 governmental awareness of wind energy mainly concentrated in Denmark and Germany, where a large number of wind turbines were manufactured and installed. Nowadays, most European governments are well aware of the potential of wind energy. Generally, the development and operation of a wind farm can be subdivided into four phases:

- Initiation and feasibility.
- Pre-building (conducted by go/no-go).
- Building.
- Operation and maintenance.

Wind energy is one of the fastest growing industries nowadays. The development in wind turbine (WT) technology is not limited to the significant increase in the size of the modern units, but also includes the high reliability and availability of the current machine. Therefore, a great competition among the manufactures established on the market and newcomers in the field is witnessed nowadays. A rapid development in the wind energy technology has made it alternative to conventional energy systems in recent years. Parallel to this development, wind energy systems (WES) have made a significant contribution to daily life in developing countries, where one third of the world's people live without electricity. Many developing nations need to expand their power systems to meet the demand in rural areas. However, extending central power systems to remote locations is too costly an option in most cases. Then, autonomous small-scale energy systems can meet the electricity demand in remote locations, even though they generate relatively little power. However, even little electricity would contribute greatly to the quality of life in some places of developing countries. Being one of the most promising autonomous power technologies, wind energy applications, in the power range from tens of Watts to kilowatts, are increasingly growing in rural areas of developing countries. Technical and economical aspects of WESs should further be improved to sustain this growth.

Chapter 2

WIND ENERGY

Since early-recorded history, people have been harnessing the energy of the wind. Wind energy propelled boats along the Nile River as early as 5000 B.C., by 200 B.C.; simple windmills in China were pumping water, while vertical-axis windmills with woven reed sails were grinding grain in Persia and the Middle East. New ways of using the energy of the wind eventually spread around the world. By the 11th century, people in the Middle East were using windmills extensively for food production; returning merchants and crusaders carried this idea back to Europe. The Dutch refined the windmill and adapted it for draining lakes and marshes in the Rhine River Delta. When settlers took this technology to the New World in the late 19th century, they began using windmills to pump water for farms and ranches; and later, to generate electricity for homes and industry.

Wind power is the conversion of wind energy into useful form, such as electricity, using wind turbines. In windmills, wind energy is directly used to crush grain or to pump water. At the end of 2007, worldwide capacity of wind-powered generators was 94.1 Giga-Watts (GW). Although wind currently produces just over 1% of worldwide electricity use, it accounts for approximately 19% of electricity production in Denmark, 9% in Spain and Portugal; and 6% in Germany and the Republic of Ireland (2007 data). Globally, wind power generation increased more than fivefold between 2000 and 2007.

The cost of the overall system increases as the complexity of the power electronic converter increases. The intricacy of the controller design also affects cost; for example, the use of MPPT techniques would cost more than a simple lookup table method. However, higher order control and converter designs may increase efficiency of the overall system. The inclusion of a DC-boost stage helps reduce the control complexity of the grid inverter at a small increase in cost. Likewise, replacing the diode rectifier with a controlled rectifier allows for a wider range of control of both the generator and grid real and reactive power transfer. In order to maximise the benefits of the wind energy conversion system, a compromise between efficiency and cost must be obtained. A summary of the different generator–converter topologies available for wind energy conversion is shown in Table 1 [9]. In addition, Table 2 lists the various generators discussed in this article and outlines the advantages and disadvantages of each [1-9].

Wind is simple air in motion. It is caused by the uneven heating of the earth's surface by the sun. Since the earth's surface is made of very different types of land and water, it absorbs the sun's heat at different rates. Today, wind energy is mainly used to generate electricity. Wind energy is also world's fastest growing energy source and is a clean and renewable source that has been in use for centuries in Europe and more recently in the United States and other nations. Wind turbines, both large and small, produce electricity for utilities and homeowners and remote villages. Wind energy is a clean energy source as electricity generated by wind turbines do not pollute the air or emit pollutants like other energy sources. This means less smog, less acid rain and fewer greenhouse gas emissions. Every 10,000 Mega Watts (MW) of wind installed can reduce carbon dioxide (CO_2) emissions by approximately 33 MMT annually if it replaces coal-fired generating capacity, or 21 million metric tonnes (MMT) if it replaces generation from average fuel mix. Many developing countries have little incentive to use wind energy technologies, to reduce their emissions despite the fact that the most rapid growth in CO_2 emissions is in the developing world. Two related activities could give both developed and developing countries incentives to develop wind projects. The first is joint implementation, a programme under which firms from the developed countries can earn carbon offsets by building clean energy projects in the developing world. Developed nations should endorse and push for joint implementation to move from its current status to full-scale implementation.

The second activity is the World Bank's Global Environmental Facility (GEF), which can cover the incremental cost of developing environmentally benign or beneficial projects in the developing world, such as building wind projects instead of an apparently cheaper coal projects. This incentive is particularly important for countries such as China and India, which have tremendous power needs and must build energy capacity quickly at the lowest possible cost. There are numerous factors that influence the overall prospects for the wind industry, though in the end, it is the economics that will be the deciding factor (Table 3). The most important issues identified:

Assessment of previous patterns of market development in similar markets.
Increased engagement of utilities and large energy companies.
National energy plans and government support for renewable energy.
Technical development.
Growth in market and the present dynamics of the industry.
Information about specific large projects.
Assessment of wind resources and how they can be used.

Economic projections are difficult at the best of times, when economies are relatively stable and a reference 'business as usual' case can be used. However, there are numerous signals that the world faces very turbulent economic conditions for a while -a credit crunch may make some projects finance difficult and the shortage of raw materials could lead to supply chain difficulties. However, the rapidly escalating price of oil is focusing a lot of attention on the price of energy and the hedge of electricity supply without a fuel cost is likely to become increasingly attractive to many companies and utilities. At some stage, rising fuel costs could lead to demand for wind energy becoming almost infinite. The main factors expected to influence the continuing growth of the wind sector are:

The economies of the transition states (Russia and Central Asia) will start to grow and increasing energy demand in Asia and South America.
Oil prices will continue to remain high as will demand for fossil fuels.
Continuing competitiveness of wind with fossil fuels.
Many countries may find they are well off their international CO_2 reduction commitments and need to install some new renewable capacity very quickly.
Security of supply questions will continue to support wind power.
Deregulated markets will remove excess conventional power capacity and new capacity is likely to be more expensive than wind.

Table 1. Summary of wind energy conversion systems

Generator (power range)	Converter Options	Device Count (Semiconductor Cost)	Control Schemes
PMSG (kW)	Diode bridge/SCR inverter compensator	DC-Link cap, 12 controllable switches (moderate)	Simple firing angle control of one converter
	SCR rectifier/SCR inverter	DC-Link cap, 12 controllable switches (moderate)	Simple firing angle control of both converters
	Diode bridge/Hard-switching inverter	DC-Link cap, 6 controllable switches (low)	Power mapping technique including stator frequency derivative control MPPT, wind prediction control
	Diode bridge/DC boost/Hard-switching inverter	DC-Link cap, 7 controllable switches (low)	Vector control of supply side inverter DC Voltage control via chopper duty radio
	Back-to-back hard-switching inverters	DC-Link cap, 12 controllable switches (moderate)	MPPT, vector control of both converters
	Back-to-back hard-switching inverters (reduced switch)	2 DC-Link cap, 8 controllable switches (low)	Generator controlled through MPPT inverter current controlled through PI controllers

DFIG (Kw-MW)	Diode bridge/SCR inverter	DC-Link cap, 6 controllable switches (low)	Sliding mode control
	SCR rectifier/SCR inverter	DC-Link cap, 12 controllable switches (moderate)	Dual thyristor firing angle control
	Back-to-back hard-switching inverters	DC-Link cap, 12 controllable switches (moderate)	Vector control of rotor and supply side space vector modulation or PWM MPPT, space vector control
	Matrix converter	18 controllable switches (high)	Vector control of rotor and supply side double space vector PWM switching
IG (kW-MW)	Back-to-back hard-switching inverters	DC-Link cap, 12 controllable switches (moderate)	Vector control, use fuzzy logic controllers user rotor slot harmonics and model reference adaptive system
SG 9kW-MW)	Diode bridge/DC boost/Hard-switching inverter	DC-Link cap, 7 controllable switches (low)	Phase angle displacement control supply voltage control
	Back-to-back hard-switching inverters	DC-Link cap, 12 controllable switches (moderate)	Supply real and reactive power control generator electromagnetic torque control

Table 2. Advantages and disadvantages of generator types

Generator Type	Advantages	Disadvantages
Permanent magnet synchronous generator	Flexibility in design allows for smaller and lighter designs	Higher initial const due to high price of magnets used
	Higher output level may be ahieved without the need to increase generator size	Permanent magnet costs restricts production of such generators for large scale grid connected turbine designs
	Lower maintenance cost and operating costs, bearings last longer	High temperatures and severe overloading and short circuit conditions can demagnetize permanent magnets
	No significant losses generated in the rotor	Use of diode rectifier in initial stage of power conversion reduces the controllability of overall system
	Generator speed can be regulated without the need for gears or gearbox	
	Very high torque can be achieved at low speeds	
	Eliminates the need for separate excitation or cooling systems	

Asynchronous generator	Lower capital cost for construction of the generator	Increase converter cost since converter must be rated at the full system power
	Known as rugged machines that have a very simple design	Results in increased losses through converter due to large converter size needed for IG
	Higher availability especially for larger scale grid connected designs	Generator requires reactive power and therefore increases cost of initial AC-DC conversion stage of converter
	Excellent damping of torque pulsation caused by sudden wind gusts	May experience a large in-rush current when first connected to the grid
	Relatively low contribution to system fault levels	Increased control complexity due to increased number of switches in converter
Double fed induction generator	Reduced converter cost, converter rating is typically 25% of total system power	Increased control complexity due to increased number of switches in converter
	Improved efficiency due to reduced losses in the power electronic converter	Stator winding is directly connected to the grid and susceptible to grid disturbances
	Suitable for high power applications including recent advances in offshore installation	Increased capital cost and need for periodic slip ring maintenance

Table 2. (Continued)

	Allows converter to generator or absorb reactive power due to DFIG used	Increased slip ring sensitivity and maintenance in offshore installations
	Control may be applied at a lower cost due to reduced converter power rating	Is not direct drive and therefore requires a maintenance intensive gearbox for connection to wind turbine
Wound field synchronous generator	Minimum machanical wear due to slow machine rotation	Typically have higher maintenance costs again in comparison to that of an IG
	Direct drive applicable further reducing cost since gearbox not needed	Magnet used which is necessary for synchronization is expensive
	Allow for reactive power control as they are self excited machines that do not require reactive power injection	Magnet tends to become demagnetized while working in the powerful magnetic fields inside the generator
	Readily accepted by eletrically isolated systems for grid connection	Requires synchronizing relay in order to properly synchornize with the grid
	Allow for independent control both real and reactive power	

Most of these factors are favourable for the industry at the moment. There is strong political support for wind energy, both as engineering and supply chain problems that have been associated with rapid growth in the past. While wind energy can still seem a small industry compared with conventional power generation, the achievement of 1% of world electricity generation is potentially significant. In individual markets such as Denmark, Germany and Spain reaching 1% has been a breakthrough figure, establishing a critical mass and being followed by further rapid growth in each year market. If the same pattern is seen with world wind energy demand and the industry continues to establish itself as a significant player in the energy sector and pushes on rapidly to 30% of world electricity demand and beyond, then the glass should be seen as half full.

Table 3. Market shares 2005-2007

Years		**2005**		**2006**		**2007**	
Manufacturer	Country	Supplied	Share	Supplied	Share	Supplied	Share
Vestas	Denmark	3186	27.6%	4239	28.2%	4503	22.8%
Ge Wind	US	2025	17.5%	2326	15.5%	3283	16.6%
Gamesa	Spain	1474	12.8%	2346	15.6%	3047	15.4%
Enercon	Germany	1640	14.2%	2316	15.4%	2769	14.0%
Suzton	India	700	6.1%	1157	7.7%	2082	10.5%
Siemens	Denmark	629	5.4%	1103	7.3%	1397	7.1%
Acciona	Spain	224	1.9%	426	2.8%	873	4.4%
Goldwind	China	132	1.1%	416	2.8%	830	4.2%
Nordex	Germany	298	2.6%	505	3.4%	676	3.4%
Sinovel	China	3	0.0%	75	0.5%	671	3.4%
Others		1032	8.9%	1094	7.3%	2076	10.5%
Total		11343	98%	16003	107%	22207	112%

Wind energy is one of the low investments high yielding sources of power generation. The future of wind energy is extremely bright and there is no doubt that in the renewable energy sector, wind power would play a predominant role in adding to the national grids clean and non-polluting energy in the coming years (Table 4).

In recent years, demand for the micro wind turbines, of the output below 1 kW, is on the increase as monuments and educational materials. Most of the micro wind turbine that has a diameter under 1.0 m is low blade tip speed ratio type on the market, by the problem of the frequency, the safety and the blade noise. In these circumstances, it would be necessary to develop the system characteristics of micro wind turbines for the purpose of much higher performance in spite of the low Reynolds number regions. Wind power generation is characterised by its stochastic nature, whereby supply and demand, in small grid systems in particular, mostly do not match. The combination of wind power with a second complementary power generation and/or direct/indirect storage technology therefore has, in principle, considerable potential. Wind-diesel, wind-water desalination and wind power in combination with hydrogen production are all potential options that have been high on the international renewable energy agenda for several years. A small-scale wind-PV hybrid power generator system for dairy farm is shown in Figure 1, to verify the possibilities to apply a power generating system and heating source for dairy farm. It is possible to apply the system for power supply and heat source to melt snow and process fertiliser.

Table 4. Installed capacity per year

Year	Europe (MW)	World (MW)
Before 2000	9.413	13.954
2000	13.306	18.449
2001	17.812	24.927
2002	23.832	32.037
2003	29.301	40.301
2004	34.725	47.912
2005	40.897	59.320
2006	48.628	74.517
2007	57.136	94.593
2008	66.785	120.458
2009	78.514	151.753
2010	93.590	191.318

Wind energy is one of the fastest growing industries nowadays. The development in wind turbine (WT) technology is not limited to the significant increase in the size of the modern units, but also includes the high reliability and availability of the current machine. Therefore, a great competition among the manufactures established on the market and newcomers in the field is witnessed nowadays. A rapid development in the wind energy technology has made it alternative to conventional energy systems in recent years. Parallel to this development, wind energy systems (WES) have made a significant contribution to daily life in developing countries, where one third of the world's people live without electricity [10].

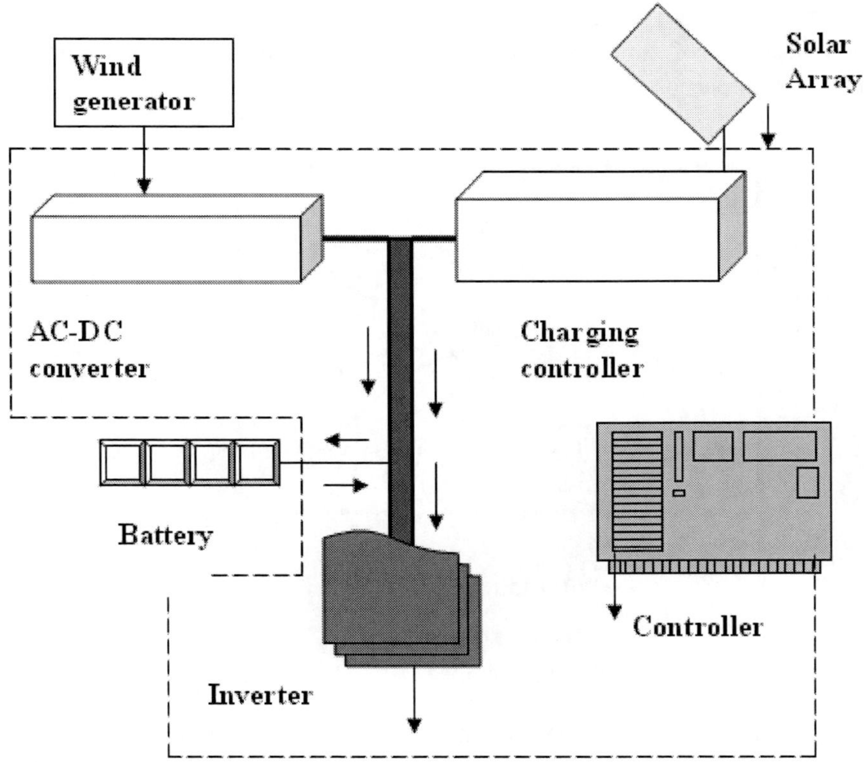

Figure 1. Wind-Photovoltaic hybrid generation systems.

Many developing nations need to expand their power systems to meet the demand in rural areas. However, extending central power systems to remote locations is too costly an option in most cases. Then, autonomous small-scale energy systems can meet the electricity demand in remote locations, even though they generate relatively little power. However, even little electricity would contribute greatly to the quality of life in some places of developing countries. Being one of the most promising autonomous power technologies, wind energy applications, in the power range from tens of Watts (W) to kilowatts (kW), are increasingly growing in rural areas of developing countries.

Technical and economical aspects of the WESs should further be improved to sustain this growth. Techno-economically optimal designs are crucial for wind systems in competing with the conventional and more reliable power systems. High performance at the lowest possible cost will encourage the use of such systems and lead to more cost effective systems gradually (Figure 2). Design tools, allowing system performance assessment over a certain period of time, are therefore of great importance for sizing and optimisation purposes.

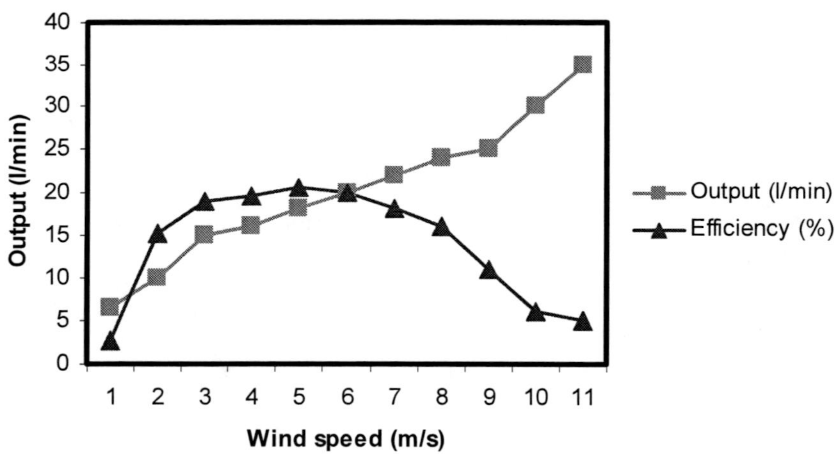

Figure 2. Performance of the wind pump.

Wind power now accounts for the dominant share of global investment in renewable energy. Total wind power capacity grew by 28% worldwide in 2007 to reach an estimated 95 GW. Annual capacity additions by market size increased even more 40% higher in 2007 compared to 2006. Wind markets have also become geographically broad, with capacity in over 70 countries. Even as turbine prices remained high, due in part to materials costs and supply-chain troubles, the industry saw an increase in manufacturing facilities in the United States, India and China, broadening the manufacturing base away from Europe with the growth of more localised supply chains. India has been exporting components and turbines for many years and it appeared that

2006 and 2007 marked a turning point for China as well, with deals announced for the export of Chinese turbines and components. The annual energy yield is calculated by multiplying the wind turbine power curve with the wind distribution function at the site:

$$E_y = \sum_{i=1}^{i=n} f_{wi} P_{wi} \tag{1}$$

where:

E_y is annual energy yield in kWh.

w is the wind speed in m/s.

n is the number of data bins converting the wind speed range of the turbine (0.5 or 1 m/s intervals).

f_{wi} is the number of hours per year for which wind speed is w m/s.

P_{wi} is the power resulting from a wind speed of w m/s.

Based on power curve from Figure 3 and the Weibull wind speed distribution, with a shape factor of 2, and the gross energy yield corresponding to 7-8.5 m/s is 10 MW.

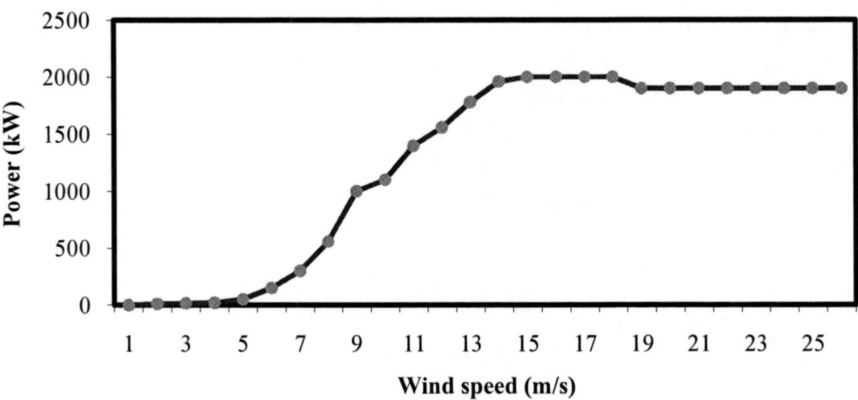

Figure 3. A power/wind speed curve.

Unchanging for all wind turbines- big or small- is a number of crucial factors that together determine the annual energy-generating potential in kWh/m^2 of rotor swept area. Key factors that impact potential energy yield and their physical relationships are expressed in the formula:

$$P = \frac{1}{2} \rho C_p \eta_{me} \eta_{el} V^3 A \quad (2)$$

where:

P is the wind turbine power performance fed into the grid (Watts).

C_p is an aerodynamic efficiency of conversion of wind power into mechanical power, often called the power coefficient.

η_{me} is the conversion efficiency of mechanical power in the rotor axis into mechanical power in the generator axis. Encompasses all combined losses in the bearings, gearbox and so on.

η_{el} is the conversion efficiency of mechanical power into electric power fed into the grid, encompassing all combined losses in the generator, frequency converter, transformer, switches, etc.

ρ is the air density in kg/m^3 depends on environmental conditions.

V is the wind speed some three-rotor diameters upwind from the rotor plane in m/s.

A is the rotor swept area in m^2.

Each of the elements of the performance formula has its own distinct contribution to total wind turbine power output and resulting yearly energy yield. Traditionally wind turbines applied in an open field are horizontal-axis designs fitted with an upwind rotor. In the operational output range, wind power generated increases with wind speed cubed. Rotor swept area is a function of the rotor diameter squared and is the second key wind turbine output variable. The Boyle-Gay-Laussac Law shows the impact of temperature and pressure on density, whereby density is proportional to pressure divided by temperature. The influence of air density on wind turbine performance is therefore limited.

2.1. Analysis and Assessment Methodologies

Three basic methods have been used in wind energy resource assessments:

Statistical and subjective analysis of existing wind measurements, other meteorological data and topographical information;

Qualitative indicators of long-term wind speed levels; and

Application of boundary layer similarity theory and the use of surface pressure observations.

In general, wind data in summarised or digitised formats are preferred. For stations having several different types of summarised wind data covering various time periods, one or two of the better summaries for those stations should be selected considering:

The most suitable format for wind power assessment;
The longest record;
The least charge in anemometer evaluation and exposure; and
The most frequent daily observations.

In many remote areas, wind data may be sparse or non-existent and evaluation of the wind data may have to rely on qualitative rather than quantitative methods. For example, there are topographic/meteorologic indicators of both high and low wind power classes. The following are some indicators of a potentially high wind power class:

Gaps, passes and gorges in areas of frequent strong pressure gradients.
Long valleys extending down from mountain ranges.
Plains and plateaus at high evaluations.
Plains and valleys with persistent down slope winds associated with strong pressure gradients.
Exposed ridges and mountain summits in areas of strong upper-air winds.
Exposed coastal sites in areas of strong upper-air winds or strong thermal pressure gradients.

Features generally indicative of low mean wind speeds are as follows:

Valleys perpendicular to the prevailing wind a lot.
Sheltered basins.
Short and/or narrow valleys and canyons.
Areas of high surface roughness (e.g., forested hilly terrain).

2.2. Statistical Distribution for Wind Data

2.2.1. Methods of Analysis

Available wind data from the Meteorological Department must be used. The data must be subsequently stratified according to quality, based on the following factors:

Accuracy of the recording equipment and techniques.
Type of data collected.
Exposure of the recording equipment.
Recording period (year).
Recording rate/interval.

2.2.2. Adjustment of Evaluation

Due to the anemometers at different meteorological stations being set at different levels, the measurements, prior to analysis, have to be adjusted to the same height. The standard height, according to the World Meteorological Organisation (WMO), is 10 meters above ground level [11]. This height is adopted in the following analysis. There are two methods that can be used to adjust the wind velocity at one level to another level. One of them is the application of power law; the other is to employ the logarithmic law.

2.2.3. Power Law

Power law is a mathematical relation representing measured wind speed profile in turbulent boundary layer. This relation is expressed in the form:

$$(V_1/V_2) = (h_1/h_2)^n \quad (3)$$

where V_1 is wind speed at height h_1 in turbulent boundary layer; V_2 is free stream wind speed; h_2 is boundary layer thickness; and n is power law exponent.

Practically, the wind speed at any height Z is adjusted to the speed at Z reference. The equation, then, becomes:

$$(V_{ref}/V_Z) = (Z_{ref}/Z)^n \quad (4)$$

or

$$V_{ref} = V_Z(Z_{ref}/Z)^n \quad (5)$$

Power law exponent varies depending on the surface roughness. It has a value of 0.14 for calm sea, 0.4 for town [12]. In 1978, Smedman-Högström and Högström [13] proposed a relationship between the exponent n and the surface roughness Z_o. Their proposed relation, deduced from the experimental results, also includes the stability of the atmosphere. This relationship can be mathematically written as:

$$n = C_o + C_1 \log Z_o + C_2 \log Z_o \quad (6)$$

Where C_o, C_1 and C_2 vary with the stability of the atmosphere, Z_o is the surface roughness length, which is shown in Tables (5) and (6).

Table 5. Roughness of height for different types of terrain

Terrain	**Types**	**Roughness height Z_o (m)**
Flat	Ocean, landscape and beach	0.005
Open	Low grass, airports, high grass and low crops	0.03
Rough	Tall row crops	0.25
Very rough	Forests	0.50
Closed	Villages	1.00
Towns	Town centre and open spaces in forests	>2.0

Table 6. Values of the constants C_o, C_1 and C_2 of the equation (4) for n as a function of log Z_o

Stability class	**C_o**	**C_1**	**C_2**
1, 2 Unstable	0.18	0.13	0.03
3 Near neutral	0.30	0.17	0.03
4 Slightly stable	0.52	0.20	0.03
5 Stable	0.80	0.25	0.03
6 Very stable	1.03	0.31	0.03

A number of effects have to be considered:

(1) Wind shear: The wind slows down, near the ground, to an extent determined by the surface roughness.
(2) Turbulence: Behind buildings, trees, ridges, etc.
(3) Acceleration: (Or retardation) on the top of hills, ridges, etc.

Wind flowing around buildings or over very rough surfaces exhibits rapid changes in speed and/or direction, called turbulence. This turbulence decreases the power output of the wind machine and can also lead to unwanted vibrations of the machine. Generally, the effect is stronger when the ridge is rather smooth and not too steep nor too flat. The orientation of the ridge should preferably by perpendicular to the prevailing wind direction. If the ridge is curved, it is best if the wind blows in the concave side of the ridge. A quantative indication of acceleration is difficult to give, but increases of 10% to 20% in wind speed are easily attained. Isolated hills give less acceleration than ridges, because the air tends to flow around the hill. This means that in some cases the two hillsides, perpendicular to the prevailing wind, are better locations than top.

The power output of wind rotor increases with the cube of the wind speed. This means that the site for a wind machine must be chosen very carefully to ensure that the location with highest wind speed in the area is selected. The site selection is rather easy in flat terrain but much more complicated in hilly or mountainous terrains. The manipulations are meant to facilitate the judgement to what extent a given location might be suitable for the utilisation of wind energy. In this respect, interest in the following:

The daily, monthly and annual wind pattern.
The duration of low wind speeds and high wind speeds.
The expected locations must be not too far from the place of measurements.
The maximum gust speed.
The wind energy produced per month and per year.

2.2.4. Logarithmic Law

Logarithmic law is the equation using physical arguments and experiment in analysis. This equation is:

$(V_Z/V_*) = 1/k \ln (Z/Z_o)$ (7)

where V_* is friction velocity; V_Z is the wind speed at height Z; k is Von Karman constant, equals to 0.4; Z_o is the surface roughness length which can be found from [14]. Monin and Obukov [15] modified equation (7) by including the stability of the atmosphere. Their modified equation becomes:

$$V_Z = V_*/k \, [\ln Z/Z_o - \Phi(Z/L)] \tag{8}$$

where Φ(Z/L) is a function with value varying with the stability of the atmosphere. For examples:

Stable condition:

$$\Phi(Z/L) = -4.7 \, Z/L \tag{9}$$

where ($1/L > 0.003 \ m^{-1}$)

Neutral condition:

$$\Phi(Z/L) = 0 \tag{10}$$

where ($-0.003 < 1/L \leq 0.003 \ m^{-1}$)

Unstable condition:

The determination of k and c was made in two steps (Wong, 1977) [16]. In the first step, the initial values are estimated by the moment method. In the second step, the maximum likelihood estimation is used to calculate the Weibull parameters (k and c are of course the solutions of the following system:

$$\delta \, Ln \, L/\delta k = 0, \ \delta \, Ln \, L/\delta c = 0$$

$$LnL = N.Ln(k) - N.k.Lnc + (k-1)\sum_{i=1}^{N} Ln(vi) - \sum_{i=1}^{N}\left(\frac{vi}{c}\right)^{k} \tag{11}$$

where N is a set of an hourly data.

Determination of the position on the earth surface for evaluating the surface roughness: Equation (7) requires the value of surface roughness. However, in order to get the value of surface roughness, it is necessary to fix the area that has an influence on wind profile at the level to be adjusted.

Smedman-Högström and Högström [13] derived the relationship between the growth of the internal boundary layer, Z_X and the distance from the discontinuity, X from the analysis of Pasquill [17]:

$$Z_X = aX^b \qquad (12)$$

In which a and b are constants which vary with stability and surface roughness.

2.2.5. Available Wind Energy

The power available (P_a) in cross sectional area A perpendicular to the wind stream moving at speed V is:

$$P_a = 0.5\ \rho\ A\ V^3 \qquad (13)$$

where ρ is the air density.
Sometime available wind energy is expressed as power density:

$$P_a/A = 0.5\ \rho\ V^3 \qquad (14)$$

However, wind machines can utilise not all of this power. The amount of power, which can be extracted from the wind stream, depends on the available wind energy and on the operating characteristics of the wind energy extraction device. The power output P of a wind energy conversion system, which subtends area A of the wind speed V, and density ρ is:

$$P = 0.5\ \eta\ C_p\ \rho\ A\ V^3 \qquad (15)$$

where η is the power coefficient (is the ratio of the actual output compared to the theoretical available = Actual power/theoretical power) [18], $C_{p\ (Betz)}$ = 16/27 = (0.593) is the theoretical maximum efficiency of the Betz Limit (in other words, theoretical maximum fraction of extracted power). This maximum is called the Betz-maximum in honour of the wind pioneer who first derived its value [19] (Appendix 1), ρ is the air density (kg m^{-3}); the density of the air depends on the temperature and on the altitude above sea level.

2.2.6. Weibull Distribution

In recent years much efforts has been made to construct an adequate statistical model for describing the wind frequency distribution. Most attention has been focused on Weibull function, since this give a good fit to the experimental data [20]. Weibull distribution is characterised by two parameters: shape parameter, K and scale parameter, C. The probability density function is given by:

$$F(V)=(K/C)(V/C)^{K-1}\ exp.[-(V/C)^{K}] \qquad (16)$$

where V is the wind speed, K is the shape parameter and C is the scale parameter.

In addition, the cumulative distribution functions by:

$$F(V) = exp.[-(V/C)^{K}] \qquad (17)$$

where V is the wind speed.

The mean of the distribution, i.e., the mean wind speed, V is equal to:

$$V = C\ \Gamma\ (1/K+1) \qquad (18)$$

where Γ is the gamma function.

Defining a reduced wind speed,

$$X = v/V \qquad (19)$$

where v is the average wind speed, and V is an accumulative wind speed parameter.

The probability density function can be rewritten as:

$$F(X) = K\ \Gamma^{K}\ (1+1/K)\ X^{K-1}\ exp.[-\Gamma^{K}\ (1+1/K)\ X^{K}] \qquad (20)$$

Moreover, the cumulative distribution functions as:

$$F(X) = 1\text{-}exp.[-\Gamma^{K}\ (1+1/K)\ X^{K}] \qquad (21)$$

There are several methods for determining the Weibull distribution parameters, for example, the method of moment, the method using the energy pattern factor, the method of maximum likelihood and the method of least square fit of the cumulative probabilities [19 and 20].

Wind is simple air in motion. It is caused by the uneven heating of the earth's surface by the sun. Since the earth's surface is made of very different types of land and water, it absorbs the sun's heat at different rates. Today, wind energy is mainly used to generate electricity. Wind energy is also world's fastest growing energy source and is a clean and renewable source that has been in use for centuries in Europe and more recently in the United States and other nations. Wind turbines, both large and small, produce electricity for utilities and homeowners and remote villages. Wind energy is a clean energy source as electricity generated by wind turbines do not pollute the air or emit pollutants like other energy sources. This means less smog, less acid rain and fewer greenhouse gas emissions (GHGs). Every 10,000 MW of wind installed can reduce CO_2 emissions by approximately 33 MMT annually if it replaces coal-fired generating capacity, or 21 MMT if it replaces generation from average fuel mix. Many developing countries have little incentive to use wind energy technologies, to reduce their emissions despite the fact that the most rapid growth in CO_2 emissions is in the developing world.

Two related activities could give both developed and developing countries incentives to develop wind projects. The first is joint implementation, a programme under which firms from the developed countries can earn carbon offsets by building clean energy projects in the developing world. Developed nations should endorse and push for joint implementation to move from its current status to full-scale implementation. The second activity is the World Bank's Global Environmental Facility (GEF), which can cover the incremental cost of developing environmentally benign or beneficial projects in the developing world, such as building a wind projects instead of an apparently cheaper coal projects. This incentive is particularly important for countries such as China and India, which have tremendous power needs and must build energy capacity quickly at the lowest possible cost. Without going into details, the materials can be ranked in terms of decreasing cost, e.g., titanium, aluminium, plastics (on average), iron and cement.

Wind energy technology and applications :

- Small, micro-generation and hybrid systems.
- Machines and wind farms.
- Offshore wind power.
- Wind resources and environmental issues.
- Connection and integration.

- National and regional programmes.
- Economic and institutional issues.

The following are concluded:

- Promoting innovation and efficient use of applicable wind energy technologies.
- Identifying the most feasible and cost effective applications of wind energy resources suitable for use.
- Highlighting the local, regional and global environmental benefits of wind energy applications.
- Ensuring the wind energy takes its proper place in the sustainable developments, supply and use of energy for greatest benefit of all, taking due account of research requirements, energy efficiency, conservation and cost criteria.
- Ensuring the financing of and institutional support for economic wind energy projects.
- Encouraging education, research and training in wind energy technology in the region.

Chapter 3

SOLAR ENERGY

Solar power is used synonymously with solar energy or more specifically to refer to the conversion of sunlight into electricity. This can be done either through the photovoltaic (PV) effect or by heating a transfer fluid to produce steam to run a generator (solar thermal). Solar energy technologies harness the sun's energy for practical ends. These technologies date from the time of the early Greeks, Native Americans and Chinese, who warmed their buildings by orienting them toward the sun. Modern solar technologies provide heating, lighting, electricity and even flight.

This study examines this emerging technology and focuses on various technical, economic, and commercial aspects of solar photovoltaics. Beginning with an overview of PV technology, including its advantages, various types of PV, and its applications, the theme goes on to explore the PV market dynamics including current and future market size, market growth and development, major trends, and barriers to the growth of PV technology. A detailed analysis and cost analysis for the commercialisation of PV technology compliments this overview. The study also includes an in-depth analysis of leading players, countries as well as companies and several case studies. "Solar Photovoltaic Market Potential" provides a comprehensive assessment of the state of solar photovoltaic technology and the potential for the market.

3.1. SOLAR PHOTOVOLTAIC

The industrial revolution spurred tremendous advances in science and technology that had the effect of exacerbating the world's energy dependence.

Over the past five decades, as the demand for energy has escalated and the consumption of fossil fuels has accelerated, people have sought renewable sources as an alternative way to meet growing energy requirements. One promising and virtually inexhaustible source of energy is the sun. Solar energy enables vegetation to grow, and it can also be used to produce electricity by way of photovoltaic systems. Photovoltaic (PV) systems convert sunlight into electricity by means of photovoltaic - or solar - cells. When sunlight shines on photovoltaic cells, it is absorbed and converted directly into electricity without any moving parts. Although each cell produces only a small amount of electricity, cells can be linked together into solar arrays until the electrical output need is met.

PV is an increasingly important energy technology. Deriving energy from the sun offers numerous environmental benefits. It is an extremely clean energy source, and few other power-generating technologies have as little environmental impact as photovoltaics. As it quietly generates electricity from light, PV produces no air pollution or hazardous waste. Moreover, it does not require liquid or gaseous fuels to be transported or combusted. In addition, because its energy source, sunlight, is free and abundant, PV systems can offer virtually guaranteed access to electric power. However, this technology faces several large obstacles, most notably the costs relating to power generation and transmission as well as difficulties in obtaining funding for the development of advanced technology. Research is underway for development of so-called second generation - or thin-film - PV technologies to bring down the costs associated with PV energy.

There is no shortage of solar-derived energy on earth and have yet to fully harness its tremendous potential to power our planet's insatiable energy demands. Since ancient times, solar energy has been harnessed for human use through a range of technologies. Photovoltaics (PV) are the use of solar cells to convert sunlight directly into electricity. Due to the growing demand for clean sources of energy, the manufacture of solar cells and photovoltaic arrays has expanded dramatically in recent years.

In the past few years, the solar PV market has experienced massive growth, in both installed capacity and production, although hindered by a global shortage of solar grade silicon. This includes opposing viewpoints within the industry regarding the anticipated period of this shortage. Globally, China has played a major role in 2006 and 2007 in the global PV market. Chinese solar PV companies have developed almost overnight and several have already conducted IPOs. A rapidly growing production capacity for solar cells and modules has been accompanied by rising production and recycling of silicon.

Most PV production in China is exported. Germany, Japan and the USA remain the world leaders, but new countries are entering the PV market as demand spreads. Recent developments have been accompanied by uncertainty due to the collapse in the statistical recording system in Germany, the largest market with half the world's installed capacity.

3.2. SOLAR THERMAL

Concentrated sunlight has been used to perform useful tasks from the time of ancient China. A legend claims Archimedes used polished shields to concentrate sunlight on the invading Roman fleet and repel them from Syracuse in 212 B.C. Leonardo Da Vinci conceived using large-scale solar concentrators to weld copper in the 15th century. In 1866, Auguste Mouchout successfully powered a steam engine with sunlight, the first known example of a concentrating solar-powered mechanical device. Over the following 50 years, inventors such as John Ericsson, and Frank Shuman developed solar-powered devices for irrigation, refrigeration and locomotion. The progeny of these early developments are the concentrating solar thermal power plants of today.

Concentrating solar thermal (CST) systems use lenses or mirrors and tracking systems to focus a large area of sunlight into a small beam. This is then used to generate electricity. Moreover, the high temperatures produced by CST systems can be used to provide process heat and steam for a variety of secondary commercial applications (cogeneration). However, CST technologies require direct insolation to function and are of limited use in locations with significant cloud cover. The main methods for producing a concentrated beam are the solar trough, solar power tower and parabolic dish; the solar bowl is more rarely used. Each concentration method is capable of producing high temperatures and high efficiencies, but they vary in the way they track the sun and focus light.

The energy conservation scenarios include rational use of energy policies in all economy sectors and use of combined heat and power systems, which are able to add to energy savings from the autonomous power plants. Electricity from renewable energy sources is by definition the environmental green product. Hence, a renewable energy certificate system is an essential basis for all policy systems, independent of the renewable energy support scheme. It is, therefore, important that all parties involved support the renewable energy certificate system in place. The potential of the most important forms of renewable energy, such as solar, wind, biomass, and geothermal energies, is shown in Tables (7 and 8). Existing renewable energy technologies could play a significant mitigating role, but the economic and political climate will have to change first. Climate change is real. It is happening now, and greenhouse gases produced by human activities are significantly contributing to it. The predicted global temperature increase of between 1.5 and 4.5 degrees C, could lead to potentially catastrophic environmental impacts. These include sea level rise, increased frequency of extreme weather events, floods, droughts, disease migration from various places and possible stalling of the Gulf Stream. This has led scientists to argue that climate change issues are not ones that politicians can afford to ignore, and policy makers tend to agree. However, reaching international agreements on climate change policies is no trivial task [21].

Table 7. Energy sources for rural areas

Source	Form
Solar energy	Solar thermal, and solar PV
Biomass energy	Woody fuels, and non woody fuels
Wind energy	Mechanical types, and electrical types
Mini and micro hydro	A mass water fall, and current flow of water
Geothermal	Hot water

Current approaches to energy are neither sustainable nor renewable. Furthermore, energy is directly related to the most critical social issues that affect sustainable development, such as poverty, jobs, income levels, access to social services, gender disparity, population growth, agricultural production, climate change and environment quality, and economic and security issues. Without adequate attention to the critical importance of energy from all of these aspects, the global social, economical and environmental goals of sustainability cannot be achieved. Indeed, the magnitude of change needed is immense, fundamental and directly related to the energy produced and consumed nationally and internationally. The key challenge to realising these targets is to overcome the lack of commitment and to develop the political will to protect people and the natural resource base [22]. Failure to take action will lead to continuing degradation of natural resources, increasing conflicts over scarce resources and widening gaps between rich and poor. Implementing sustainable energy strategies is one of the most important levers humankind has for creating a sustainable world. More than 2 billion people have no access to modern energy sources, most of them are living in rural areas. Food and fodder availability is very closely related to energy availability [22].

Table 8. Potential, productive, end-uses of various energy sources and technologies

Energy source and technology	Productive end-uses and commercial activities
Solar	Lighting, water pumping, radio, TV, battery charging, refrigerators, cookers, dryers, cold stores for vegetables and fruits, water desalination, heaters, baking, etc.
Wind	Pumping water, grinding and provision for power for small industries
Hydro	Lighting, battery charging, food processing, irrigation, heating, cooling, cooking, etc.
Biomass	Sugar processing, food processing, water pumping, domestic use, power machinery, weaving, harvesting, sowing, etc.
Kerosene	Lighting, ignition fires, cooking, etc.
Dry cell batteries	Lighting, and small appliances
Diesel	Water pumping, irrigation, lighting, food processing, electricity generation, battery charging, etc.
Animal and human power	Transport, land preparation for farming, and food preparation (threshing)

In order to meet challenges, the future energy policies should put more emphasis on developing the potential of energy sources, which should form the foundation of future global energy structure [22]. The concept of an integrated renewable energy farm (IREF) is a farming system model with optional energetic autonomy, which includes food production and if possible, energy exports. Energy production and consumption at the IREF have to be environmentally friendly, sustainable and eventually based mainly on renewable energy sources. A combination of different possibilities exists for non-polluting energy production, such as modern wind and solar electricity production, as well as the production of energy from biomass. An IREF system based largely on renewable energy sources would seek to optimise energetic autonomy and an ecologically semi-closed system while also providing socio-economic viability and giving due consideration to the newest concept of landscape and bio-diversity management. Ideally, it would promote the integration of different renewable energies and rural development, as well as contributing to the reduction of greenhouse gas emission as shown in Table 9. The overall objective is that the IREF concept be successfully introduced into agricultural production systems, which have to be completely sustainable, taking into account the following influential factors [22]:

Impact, influence and needs of climate, soil and crops.
Ratio of required food/bio-fuel production.
Input/output requirement for cultivation, energy balance and output/input ratio.
Equipment choices (wind, solar, biomass generation and conversion technology).

Two of the most essential natural resources for all life on the earth and for man's survival are sunlight and water. Sunlight is the driving force behind many of the renewable energy technologies. The worldwide potential for utilising this resource, both directly by means of the solar technologies and indirectly by means of biofuels, wind and hydro technologies is vast. During the last decade, interest has been refocused on renewable energy sources due to the increasing prices and fore-seeable exhaustion of presently used commercial energy sources. Renewable sources of energy are regional and site specific. It has to be integrated in the regional development plans. Solar power is used synonymously with solar energy or more specifically to refer to the conversion of sunlight into electricity. This can be done either through the photovoltaic effect or by heating a transfer fluid to produce steam to run a generator. Solar energy technologies harness the sun's energy for practical ends. These technologies date from the time of the early Greeks, Native Americans and Chinese, who warmed their buildings by orienting them toward the sun. Modern solar technologies provide heating, lighting, electricity and even flight.

Table 9. The possible shares of different renewable energies in diverse climatic zones

Climatic Region	Energy source	Power production (% of total need)	Heat production (% of the total need)	Biomass need (total area)	Biomass area (5 of the total area)
Northern and Central Europe	Solar 200 m^2 Wind 100 kW Biomass	7 100 100	15 - 105	60	12
South Europe	Solar 250 m^2 Wind 100 kW Biomass	12.7 100 70	40 - 65	36	48
Northern Africa Sahara	Solar 300 m^2 Wind 100 kW Biomass	21 75 25	90 - 29	14	1.2
Equatorial Region	Solar 200 m^2 Wind 100 kW Biomass	18.2 45 70	37.5 - 80	45	N.A.

Chapter 4

BIOMASS ENERGY

Biofuels are emerging in a world increasingly concerned by the converging global problems of rising energy demands, accelerating climate change, high priced fossil fuels, soil degradation, water scarcity and loss of biodiversity. For instance, the Intergovernmental Panel on Climate Change (IPCC) identified that in order to avoid more than an acceptable maximum 2-2.4°C rise in mean global temperature, greenhouse gas emissions will need to peak around 2015 and be reduced well below 50% of 2000 level by 2050. A lower figure is needed, which cannot be achieved by emissions reduction alone. Hence, there is a need for carbon removals, giving rise to enhanced supplies of biomass raw material and the potential of biofuels-related investments to show a profit from biofuels sales revenues.

Many nations have the ability to produce their own biofuels derived both from agricultural and forest biomass and from urban wastes, subject to adequate capacity building, technology transfer and access to finance. Trade in biofuels surplus to local requirements can thus open up new markets and stimulate the investment needed to promote the full potential of many impoverished countries. Such a development also responds to the growing threat of passing a tipping point in climate system dynamics. The urgency and the scale of the problem are such that the capital investment requirements are massive and more typical of the energy sector than the land use sectors. The time line for action is decades, not centuries, to partially shift from fossil carbon to sustainable biomass. Figure 4 shows the contribution of biomass to global primary and consumer energy supplies. Food and fodder availability is very closely related to energy availability.

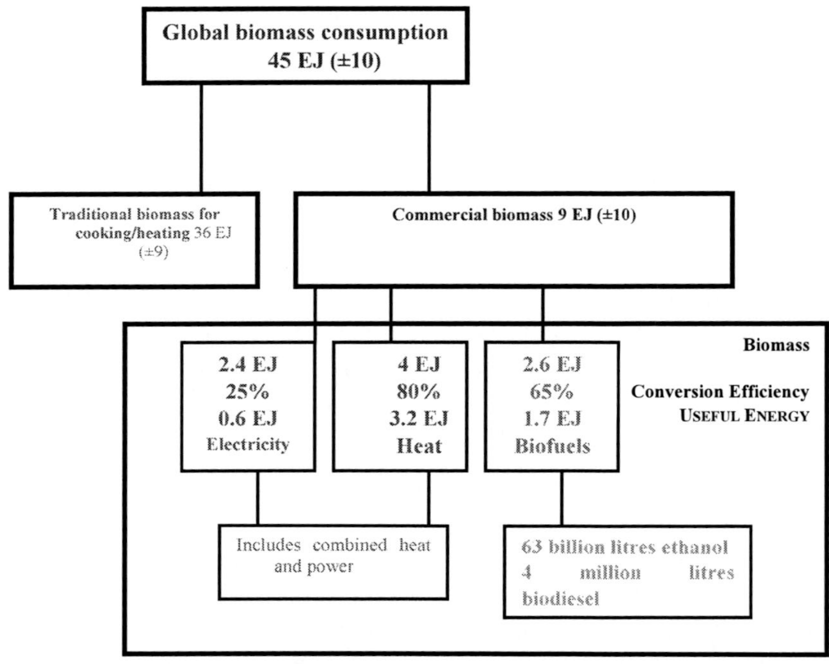

Figure 4. Contribution of biomass to global primary and consumer energy supplies.

4.1. Biofuels

Biofuel is any fuel that is derived from biomass - recently living organisms or their metabolic by-products, such as manure from cows. It is a renewable energy source, unlike other natural resources such as petroleum, coal, and nuclear fuels. Ethanol is manufactured from microbial conversion of biomass materials through fermentation. Ethanol contains 35% oxygen. The production process consists of conversion of biomass to fermentable sugars, fermentation of sugars to ethanol, and the separation and purification of the ethanol. Fermentation initially produces ethanol containing a substantial amount of water. Distillation removes the majority of water to yield about 95% purity ethanol, the balance being water. This mixture is called hydrous ethanol. If the remaining water is removed in a further process, the ethanol is called anhydrous ethanol and is suitable for blending into gasoline. Ethanol is "denatured" prior to leaving the plant to make it unfit for human consumption by addition of a small amount of products such as gasoline.

Biodiesel fuels are oxygenated organic compounds - methyl or ethyl esters - derived from a variety of renewable sources such as vegetable oil, animal fat, and cooking oil. The oxygen contained in biodiesel makes it unstable and requires stabilisation to avoid storage problems. Rapeseed methyl ester (RME) diesel, derived from rapeseed oil, is the most common biodiesel fuel available in Europe. In the United States, biodiesel from soybean oil, called soy methyl ester diesel, is the most common biodiesel. Collectively, these fuels are referred to as fatty acid methyl esters (FAME). Biofuels have become a growth industry with worldwide production more than doubling in the last five years. The rapid expansion of ethanol production in the United States and biodiesel production (and to a lesser extent, biogas) in Germany and other countries in Western Europe has created a biofuels frenzy that has affected many countries, including Canada. Many measures have been used to stimulate production and consumption of biofuels, including preferential taxation, subsidies, import tariffs and consumption mandates. Recently, Canadian federal and provincial governments have announced consumption mandates and subsidies to assist rapid expansion of biofuel production in Canada. Canada has considerable natural resources and is one of the world's largest producers and exporters of energy. In 2006, Canada produced 21.1 quadrillion British Thermal Units (Btu) of total energy, the fifth largest amount in the world. Since 1980, Canada's total energy production has increased by 86%, while its total energy consumption has increased by only 48% during that period. Almost all of Canada's energy exports go to the United States, making it the largest foreign source of the USA energy imports. Canada is consistently among the top sources for the USA oil imports, and it is the largest source of the USA natural gas and electricity imports. Recognising the importance of the energy trade between the two countries, both participate in the North American Energy Working Group, which seeks to improve energy integration and cooperation between Canada, the USA, and Mexico. In the European Union (EU), transport is responsible for an estimated 21% of all greenhouse gas emissions that are contributing to global warming and this percentage is rising. In order to meet sustainability goals, in particular the reduction of greenhouse gas emissions agreed under the Kyoto Protocol, it is therefore essential to find ways of reducing emissions from transport. In light of this objective, along with diversifying fuel supply sources and developing long-term replacements for fossil oil, the European Commission proposed targets for biofuels in transport fuel by 2020 among the member states. This binding is a part of long-term energy package, which includes an overall binding 20% target for renewable energy. Under this, each member state will have to establish National Action Plans for their specific objectives and sectoral targets.

Biofuels have been produced on an industrial scale in the Europe since the 1990s but production significantly accelerated starting in the early 2000s, largely in response to rising petroleum prices and favourable legislation passed by the EU institutions and member states. Biofuels have been promoted as part of the EU strategy to encourage renewable energy and their production and use has expanded rapidly. Although EU measures have applied equally, most of the time, to biodiesel and ethanol, biodiesel production has developed at a faster rate. Biodiesel accounts for 80% of European biofuels production and ethanol for the remaining 20%. European Union is by far the biggest producer of biodiesel in the world and the reason for the big share of biodiesel is that the majority of the cars in the EU are diesel cars and as such, there is a diesel deficit. The most important feedstock for the EU biodiesel is rapeseed. Despite producing a significant portion of global biodiesel and increasing production of biofuel for transport, the EU faces a number of significant challenges in the coming years. Most important is the limited availability of land to cultivate biodiesel input crops such as rapeseed, although Ukraine's EU accession could help alleviate this constraint. A further challenge is that even with the use of the most advanced production technologies, biofuels produced in the EU are not cost competitive with fossil fuels at current oil price levels. New input crops and production methods could make biofuels more competitive.

4.2. Microwave Power Transmission

Microwave power transmission (MPT) is the use of microwaves to transmit power through outer space or the atmosphere without the need for wires. It is a sub-type of the more general wireless energy transfer methods, and is the most interesting because microwave devices offer the highest efficiency of conversion between DC electricity and microwave radiative power. Following World War II, which saw the development of high-power microwave emitters known as cavity magnetrons, the idea of using microwaves to transmit power was researched. In 1964, William C. Brown demonstrated a miniature helicopter equipped with a combination antenna and rectifier device called a rectenna. The rectenna converted microwave power into electricity, allowing the helicopter to fly. In principle, the rectenna is capable of very high conversion efficiencies - over 90% in optimal circumstances. The common reaction to microwave transmission is one of concern, as microwaves are generally perceived by the public as dangerous forms of radiation - stemming from the fact that they are used in microwave ovens. While high power microwaves can be painful and dangerous as in the United States Military's Active Denial System, MPT systems are generally proposed to have only low intensity at the rectenna. MPT is the most commonly proposed method for transferring energy to the surface of the earth from solar power satellites or other in-orbit power sources.

4.3. Ethanol Production

In 2003, European Commission issued directives that will govern European biofuels policy through 2010 and target of 5.75% biofuels consumption in the transportation sector by 2010. These include measures to increase ethanol demand and supply and providing tax benefits and exemptions to facilitate growth. The principal goals propelling bioethanol in the European countries are improving energy security, boosting rural development, and reducing greenhouse emission reductions. Transport is responsible for approximately 21% of the EU's greenhouse gas emissions, and recent the European Commission directives have made biofuels in transport a regional priority. Not all member states are equally committed to the objectives set by the European Commission, but all are trying to some extent to achieve the EU targets. Biodiesel accounts for 80% of the European biofuels production and ethanol for the remaining 20%. Active market actors and lobbying groups have contributed immensely to the evolution of the market in recent years.

However, some issues are concerning the overall growth of the ethanol industry in Europe. Most import among them has been the recent rise in the prices of food grains and subsequent decline in their supply. Moreover, many for this crisis have blamed biofuels, including ethanol. However, the supporters of ethanol and other biofuels suggest that the global food crises are a result of growing oil prices along with increased food consumption in developing world and declining yields of food crops. Another issue affecting further expansion of ethanol in Europe has been smaller number of cars in Europe that can run on ethanol. Since a majority of the cars in the EU are diesel cars and there has been a diesel deficit, the focus has been on biodiesel. However, bioethanol has the advantage over biodiesel that it can be produced from a much larger variety of different feedstock. Furthermore, the ethanol industry in the EU has also had problems to compete with cheap imports of bioethanol, especially from Brazil. These cheap imports have made it very difficult for the local industry to grow strong and manage without subsidies. Most of the cheap imports have come through a loophole in Sweden. This loophole was closed in January 2006, and there are now enormous amounts of planned bioethanol production plants in the EU. A further challenge is that even with the use of the most advanced production technologies, bioethanol produced in the EU are not cost competitive with fossil fuels. According to the most recent estimates, European ethanol would only break even at an oil price of ($115) per barrel. New input crops and production methods could make ethanol more competitive. Lignocellulosic processing and biomass-to-liquid technologies have been mentioned as potential lower-cost alternatives to current technologies. Countries including Germany and the UK are actively promoting research into second-generation biofuels.

Chapter 5

CLIMATE CHANGE, GLOBAL WARMING AND THE ENHANCED GREENHOUSE EFFECT

Industry's use of fossil fuels has been blamed for our warming climate. When coal, gas and oil are burnt, they release harmful gases, which trap heat in the atmosphere and cause global warming. However, there has been an ongoing debate on this subject, as scientists have struggled to distinguish between changes, which are human induced, and those, which could be put down to natural climate variability. Industrialised countries have the highest emission levels, and must shoulder the greatest responsibility for global warming. However, action must also be taken by developing countries to avoid future increases in emission levels as their economies develop and population grows. Human activities that emit carbon dioxide (CO_2), the most significant contributor to potential climate change, occur primarily from fossil fuel production. Consequently, efforts to control CO_2 emissions could have serious, negative consequences for economic growth, employment, investment, trade and the standard living of individuals everywhere. Scientifically, it is difficult to predict the relation between global temperature and greenhouse gas concentrations. The climate system contains many processes that will change if warming occurs. Critical processes include heat transfer by winds and currents, the hydrological cycle involving evaporation, precipitation, runoff and groundwater and the formation of clouds, snow, and ice, all of which display enormous natural variability. The equipment and infrastructure for energy supply and use are designed with long lifetimes, and the premature turnover of capital stock involves significant costs. Economic benefits occur if capital stock is replaced with more efficient equipment in step with its normal replacement cycle. Likewise, if opportunities to reduce future emissions are taken in a timely manner, they should be less costly. Such flexible approaches would allow society to take account of evolving scientific and technological knowledge, and to gain experience in designing policies to address climate change. Mitigation measures that could be under-taken to influence the effect of the oil industry that may contribute to decrease greenhouse gases (GHGs) emissions and decelerate the threat of global climate change may include the following:

Controlling GHG emissions by improving the efficiency of energy use, changing equipment and operating procedures.

Controlling GHG emission detection techniques in oil production, transportation and refining processes.

More efficient use of energy-intensive materials and changes in consumption patterns.

A shift to low carbon fuels, especially in designing new refineries.

The development of alternative energy sources (e.g., biomass, solar, wind, hydro-electrical and cogeneration).

The development of effective environment standards, policies, laws and regulations, particularly in the field of oil industry.

Embarking on conservation energy and reduction of pollution of the environment should be undertaken without delay, as a preventive measure against shortage of future energy supply against prospective national energy demand. Saving on fossil fuel for premium users/export and accelerating development of new and/or remote lands otherwise deprived of conventional energy sources are examples of such measures. Other measures should include the followings:

Launching public awareness campaigns among investor's, particularly small-scale entrepreneurs and end users of renewable energy technologies, to highlight the importance and benefits of renewables.

To direct resources away from feeding wars and the arms industry towards real development, this will serve the noble ends of peace and progress.

The energy crisis is a national issue and not only a concern of the energy sector. Any country has to learn to live with the crisis for a long period, and develop policies, institutions and manpower for more effective and long-term solutions.

To invest in research and development through the existing specialised bodies.

To encourage co-operation between nations. This should be much easier in this era of information and communications.

The government should give incentives to encourage the household sector to use renewable energy technologies instead of conventional energy.

Promotion of research and development, and demonstration and adaptation of renewable energy resources (solar, wind, biomass, mini-hydro, etc.) amongst national, regional, and international organisations, which seek clean, safe, and abundant energy sources.

Execute joint investments between the private-sector and the financing entities to disseminate the renewables with technical support from the research and development entities.

Promotion and general acceptance of renewable energy strategies by supporting comprehensive economic energy analysis taking account of environmental benefits.

Climate change is one of the most serious concerns of the society. The gases causing the greenhouse effect are carbon dioxide (CO_2), methane (CH_4), nitrous oxide (NO_x), chlorofluorocarbons (CFC's), tropospheric ozone, and stratospheric water vapour. Carbon dioxide, in particular, is an important greenhouse gas and a major agent of climate change. It is the most significant greenhouse gas, contributing about half of the total greenhouse effect [23]. The recent concentration of atmospheric CO_2 is 25% higher than the preindustrial level [23]. This increase is primarily due to fossil fuel combustion and deforestation. Climate change might pose a serious threat to our ecological and socio-economic systems unless measures are investigated to mitigate the rising accumulation of atmospheric CO_2. Increasing concern about climate change has recently generated studies about the effects of urban greenspace on the reduction of atmospheric carbon; C. Greenspace in urban ecosystems can reduce atmospheric carbon levels in three ways. In this context, greenspace is defined as soil surface area capable of supporting vegetation and the vegetation being supported. First, urban trees and shrubs directly sequester and accumulate atmospheric carbon in the process of their growth through photosynthesis. Second, urban vegetation decreases building cooling demand by shading and evapotranspiration, and heating demand by windspeed reduction, thereby reducing carbon emissions associated with fossil fuel use. Third, urban soils store organic carbon from litter-fall, until it is returned to the atmosphere by decomposition. On the other hand, buildings, factories, and automobiles in urban landscapes release carbon through fossil fuel consumption. Economic of environmental issues are increasing. However, new technologies are expected to reduce pollution derived both from productive processes and products.

The reduction of atmospheric carbon by urban greenspace could be significantly variable both within and among nations due to differences in land use distributions, greenspace sizes and structures, and fossil fuel uses. Diverse regional or national studies will help increase understanding of urban greenspace effects on atmospheric carbon levels. Urban greenspace makes only a partial contribution to atmospheric carbon reduction. However, urban greenspace planning and management could be one of more timesaving and cost-effective ways to slow climate change as compared to the development of alternative energy sources. To increase carbon storage and uptake by urban greenspace, the following strategies are suggested [23].

First, hard surfaces were predominant in the urban lands. Locating parking lots, above ground utility lines below ground could increase available growing space, and planting potential.

Second, plantings in urban lands were characterised by a single layer of grass, shrubs, or trees and a young even-aged tree population. Multi-layered plantings, in which herb, shrub, and tree layers overlap, would increase the carbon storage and uptake per unit area, and maintain a multi-aged tree structure to promote continuous carbon uptake over time.

Third, ordinances should be amended to increase greenspace area in urban lands. Fourth, tree species that have high growth rates, even under adverse urban growing conditions, should be planted. Broad-leaved tree species have higher growth rates than conifers and shrubs.

Finally, natural lands, in which carbon storage and uptake per unit area are significant, should be conserved. Disturbing natural lands will release the carbon stored in vegetation and soils back to the atmosphere. Vegetation will fix atmospheric carbon during its growing period. However, the entire carbon stored in vegetation will ultimately be lost upon the death or removal of the vegetation. Thus, vegetation is not a permanent carbon sink. Immediate replacement is necessary to compensate for the carbon emitted from previously removed vegetation. The appropriate plantings and maintenance of trees are needed to ensure longer productive life spans and to avoid rapid carbon release. More detailed studies, including annual carbon fluxes for grass and soils, would improve the understanding of greenspace impacts on carbon cycling in urban lands. Renewable energy resources are expected to play a key role in meeting the world's energy demand over the coming decades [23]. Unfortunately, these resources are all susceptible to variations in climate, and, hence, vulnerable to climate change. Recent findings in the atmospheric science literature suggest that the impacts of greenhouse gas including warming are likely to significantly alter climate patterns in future.

The flux of solar radiation reaching the surface of the earth is the primary source among all types of known renewable energies. The solar radiation data may be considered as an essential requirement for the design of the solar systems and many other industrial and agricultural projects. Thermal comfort requires constant indoor air temperature and humidity during the day. In summer, the outside air temperature usually exceeds the inside temperature and a heat flux from the ambient to the interior of the building occurs. There is therefore a need for a climatisation system to keep the comfort parameters constant. On the other hand, the decrease in thermal comfort is normally associated with time periods of abundant solar energy. The productivity and efficiency of most forms of renewable energy are constrained by environmental conditions. Solar energy, whether in the form of solar thermal energy conversion or photovoltaic (PV) energy generation, is vulnerable to variations in cloud cover and atmospheric turbidity. Likewise, the availability of biomass stocks for use as bio-fuels depends upon growing conditions such as temperature, precipitation, and incident solar radiation. While wind power generation is also susceptible to variations in ambient temperatures, humidity, and precipitation the primary determinals of available wind power are wind speed statistics (e.g., mean wind speeds and gustiness). Decadal and multi-decadal variability in wind speed statistics currently introduce an element of risk into the decision process for citing new wind power generation facilities. Recent findings from the atmospheric science community suggest that the climate change may introduce and added risk in this process. For example, carbon dioxide emissions from fossil fuel consumption are contributing to global warming and are projected to have dramatic impacts on global climate on decade to century timescales. These impacts will affect the statistics (maximum, minimum, mean and variance) of all meteorological variables. Previously, most of the interest in climate change has been focused on changes in temperature and precipitation. Most of the climate change scenarios investigated is based on the assumption that greenhouse gases will increase at a rate of 1% per year, doubling after about 70 years [24]. Results from General Circulation Models (GCM) of the climate suggest that the warming influence of this increase in carbon dioxide combined with the cooling influence of increased atmospheric aerosols will result in globally averaged annual temperature increases in the range of 2.4°C [24].

The climate system is known to have significant natural variability. This natural variability, in fact, has made it difficult to clearly identify the greenhouse gas-warming signal in current observational records. Analyses of past variations in climate, however, have identified natural variability cycles that can be subtracted from the climate record to reveal climate anomalies associated with greenhouse warming. Global warning is due to atmospheric changes and anthropogenic heating and may be exacerbated by an increasing natural variability trend over the coming decades. There is increasing evidence in the atmospheric science community that greenhouse gas concentrations will continue to increase in the future and that this will result in significant climate warming and changes in the other climate variables [24].

5.1. Environmental Policies and Industrial Competitives

The industrial development strategy of worldwide gives priority to the rehabilitation of the major industrial areas with respect to the improvement of their infrastructure in terms of roads, water supply, power supply, sewer systems, etc. This strategy takes into consideration the importance of incorporating the environmental dimension into these plans. However, the relationship between environmental policies and industrial competitiveness has not been adequately examined [25]. For the near future, the real issue concerns the effectiveness of environmental expenditure in terms of reduction of pollution emissions per unit of output.

5.2. Energy in Buildings

Energy efficiency brings health, productivity, safety, comfort and savings to homeowner, as well as local and global environmental benefits. The use of renewable energy resources could play an important role in this context, especially with regard to responsible and sustainable development. It represents an excellent opportunity to offer a higher standard of living to local people and will save local and regional resources. Implementation of greenhouses offers a chance for maintenance and repair services. It is expected that the pace of implementation will increase and the quality of work to improve in addition to building the capacity of the private and district staff in contracting procedures. The financial accountability is important and should be made transparent [26-27].

Sustainability has been defined as the extent to which progress and development should meet the need of the present without compromising the ability of the future generations to meet their own needs [28]. This encompasses a variety of levels and scales ranging from economic development and agriculture, to the management of human settlements and building practices. This general definition was further developed to include sustainable building practices and management of human settlements. The following issues were addressed during the Rio Earth Summit in 1992 [29]:

The use of local materials and indigenous building sources.
Incentive to promote the continuation of traditional techniques, with regional resources and self-help strategies.
Regulation of energy-efficient design principles.
International information exchange on all aspects of construction related to the environment, among architects and contractors, particularly non-conventional resources.
Exploration of methods to encourage and facilitate the recycling and reuse of building materials, especially those requiring intensive energy use during manufacturing, and the use of clean technologies.

The objectives of the sustainable building practices aim to:

Develop a comprehensive definition of sustainability that includes socio-cultural, bio-climate, and technological aspects.
Establish guidelines for future sustainable architecture.
Predict the CO_2 emissions in buildings.
The proper architectural measure for sustainability is efficient, energy use, waste control, population growth, carrying capacity, and resource efficiency.
Establish methods of design that conserve energy and natural resources.

A building inevitably consumes materials and energy resources. The technology is available to use methods and materials that reduce the environmental impacts, increase operating efficiency, and increase durability of buildings (Table 10).

Table 10. Design, construction and environmental control description of traditional and new houses

Design characteristics	Traditional houses	New houses
Form	Courtyard (height twice its width) - open to sky	Rectangle-closed
Construction	Brick walls 50 cm thick, brick roof with no insulation in either	Brick walls 25 cm thick, concrete roof-no insulation
Environmental control	Evaporative air coolers	Evaporative air coolers
Ease of climatic control	Difficult-rooms open into the climatically uncontrolled open courtyard	Moderate-rooms open into the enclosed internal corridor
Maintenance	Well conserved and maintained	New construction
Windows	Vertical and single glazing	Horizontal-single glazing
Urban morphology	Each house attached from 3 sides	Row houses attached from 2 sides
Orientation	Varies-irregular shapes and winding alleyways	North-south (row houses)
Orientation and solar gain	Solar gains less affected by orientation due to shading provided by the deep courtyard	Solar gain determined by orientation-no obstruction (shading)
Sharing solar gain	Significant - due to long-wave exchange or convective exchange between the 4 vertical walls surrounding the courtyard	Minimal- S-wall receives much more solar radiation than the N-wall due to the absence of any interreflection and long-wave exchange
Occupant's social status	Low income families	Low and middle income families

The comfort in a building depends on many environmental parameters. These include temperature, relative humidity, air quality and lighting. Although greenhouse and conservatory originally both meant a place to house or conserve greens (variegated hollies, cirrus, myrtles and oleanders), a greenhouse today implies a place in which plants are raised while conservatory usually describes a glazed room where plants may or may not play a significant role. Indeed, a greenhouse can be used for so many different purposes. It is, therefore, difficult to decide how to group the information about the plants that can be grown inside it.

Chapter 6

NATURAL DISASTERS

The following is a detailed guide for communities and emergency operations team to develop and maintain a viable disaster management and recovery plan. In addition, it explains in details the concept of disaster management and various steps from planning to prepare against any disaster. It further highlights the role of government agencies and local authorities at the time of disaster as well as before and after it. It also discusses long and short-term goals for mitigation, planning and recovery from disaster along with initiatives that the USA government has taken in recent years. The theme has a special focus on management of energy infrastructure during the time of disaster and presents a checklist for emergency response and recovery (Figure 5).

Every year, tornadoes, hurricanes and other natural disasters injure and kill thousands of people and damage billions of dollars worth of property in the United States. Most of the times, it is almost impossible to prevent the occurrence of these disasters and their damages. However, it is possible to reduce their impact by adopting suitable disaster management strategy. Disaster management is a systematic approach towards preparing for disaster before it happens and includes disaster response - emergency evacuation, quarantine, mass recontamination - as well as supporting and rebuilding society after natural disasters have occurred. Efficient disaster management relies on thorough integration of emergency plans at all levels of government and non-government involvement.

Disaster preparedness, emergency management and post disaster recovery is highly dependent on economic and social conditions local to the disaster. However, the basics steps for disaster management remain same in all scenarios. Preparedness is the first step to counter disaster, which involves developing plan of action, and it includes communication, chain of command development, proper maintenance and training of emergency services and development of emergency warning systems along with emergency shelters and evacuation plans. Next step is response, which includes mobilisation of the necessary emergency services such as fire-fighters, police, and ambulance that may be supported by a number of secondary emergency services, such as specialist rescue teams. Recovery from disaster involves restoration of the affected area including destroyed property, re-employment and redevelopment of essential infrastructure. Mitigation efforts attempt at preventing hazards from developing into disasters or reducing the impact of disasters and it focuses on long-term measures for reducing or eliminating future risks.

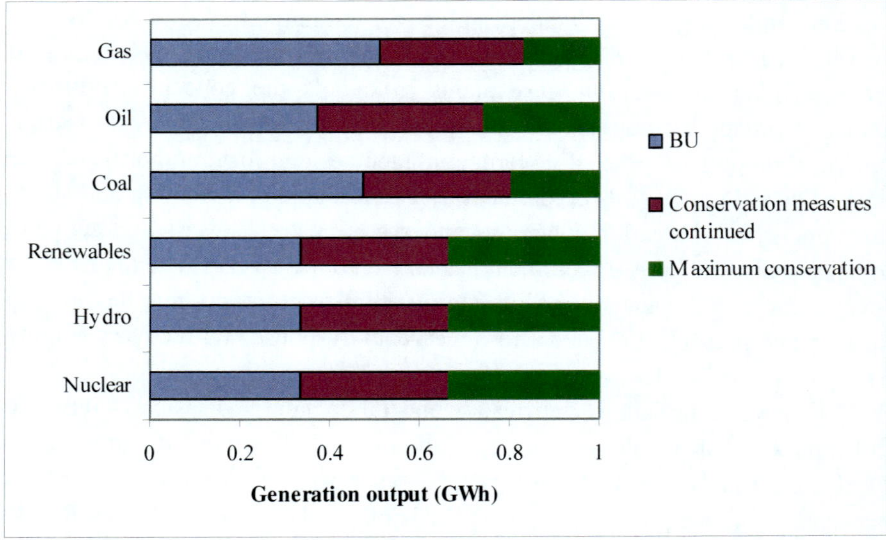

Figure 5. Energy conservation measures.

Chapter 7

WATER RESOURCES

Water is essential for life and for most activities of human society. Both economic and social development and the maintenance of human health are completely dependent upon ready access to adequate water supplies. All societies require water both for basic survival and for economic development. The indicator of naturally available water resources per capita has become the standard index for measuring the degree to which a country is facing water scarcity and is often used to show a growing global water crisis. Problems satisfying water resource needs or demands are affecting a growing proportion of the world, primarily in arid and semi-arid regions where population pressures are considerable and demand for water is currently rising faster than at anytime previously [30]. The relatively comprehensive development indicator database of the United Nations Development Programme (UNDP) and the water resources database of the Food and Agricultural Organisation (FAO) provide no support for the notion that the naturally available water resources of a country have a significant effect on the ability of that country to meet the basic needs of its population. Where water supply costs are high, and the water sector will directly contribute more to a nation's growth domestic product (GDP) compared to where costs are low because the sector will be more significant economically. With increasing globalisation of trade, global water interdependencies and overseas externalities are likely to increase. At the same time, liberalisation of trade creates opportunities to increase physical water savings. International water dependencies are substantial and are likely to increase with continued global trade liberalisation.

Chapter 8

PROBLEMS AND DIFFICULTIES IN RURAL ENERGY DEVELOPMENT

The following problems are summarised:

8.1. IMBALANCE IN RURAL ENERGY DEVELOPMENT

Due to the difference in economic conditions in different areas, the development of rural energy is considerably imbalanced. The main challenge to energy policymakers in the 21st century is how to develop and manage adequate, affordable and reliable energy services in a sustainable manner to fuel social and economic development. Generally, future rural energy will be oriented towards green energy, and the future development of rural energy will concentrate on biogas, small hydropower, solar energy, and wind power.

8.2. INSUFFICIENT INVESTMENT IN DEVELOPMENT OF RURAL ENERGY

Current rural energy relies mainly on charcoal, firewood and green energies, such as electricity and biogas. The bad economic situation leads to considerable difficulty in the development of rural energy, and farmers in remote areas still prefer ''free firewood'' for their cooking due to their low income. Hence, further development of rural energy needs significant financial support from the government at various levels. Although the work of rebuilding traditional stoves has been almost finished, most of the rebuilt fuel-saving stoves have a thermal efficiency less than 20%.

8.3. Excessive Dependence on Forests for Rural Energy

Currently, energy for rural household use comes mainly from burning of firewood. The annual consumption of forests is 1.96 x 10^6 m^3, and of this 0.65 x 10^6 m^3 is as firewood. To some extent, this pattern of energy consumption has led to environmental damage such as water and soil loss, decrease in forest cover, and air pollution. The excessive use of firewood from forests for rural energy would cause damage to sightseeing resorts, make animals lose their habitats, and lead to the extinction of some endangered plants. The future development of rural energy should be aimed at completely changing the current pattern of energy consumption, fully utilising abundant resources of hydropower, biomass, solar and wind energy, promoting economic growth through the development of rural energy and integrated utilisation of biomass.

8.4. Refrigeration

The refrigeration industry has had to face environmental challenges. Regulation and behaviour related to the ozone layer differ from one country to another. Problems related to global warming are likely to cause similar problems in very different ways. These two kinds of problems may lead company managers to change/modernise equipment. These environmental factors must not hide other important development factors such as those related to hygiene, organoleptic quality, control methods and equipment packaging, etc. All equipment change and investment projects must take all of them into consideration.

The more developed a country, is the more widely refrigeration is used. Food preservation is still the main use of refrigeration, followed by air conditioning, energy savings and transport (heat pump and liquefied gases), industrial processes and medicine. When consumers demand more and more fresh products, handling such products becomes increasingly difficult: cold stores, display cabinets, and refrigerated trucks have to be much more effective. Difficulties are mostly related to:

Very narrow ranges, in many cases, between temperatures involving microbial or chemical risk and temperatures that cause chilling injury or freezing.

No or little overlapping between temperature ranges permitted for different products.

No or little possibility for the consumer or the seller to evaluate the remaining life spans of the product or even to perceive any risk.

Humidity control, and vapour pressure differences have serious consequences, not only on the weight but also on the unit value of products.

The production and use of CFC refrigerants have been phased out in developed countries. Hydro fluorocarbons (HCFCs) refrigerants are now subject to regulation and are scheduled to be phased out in the twenty-first century. Recent developments in refrigeration technology and environment (climate) control have provided a better quality of life for mankind. As is known, refrigeration technology has played a key role in preserving the quality of food and eliminating waste (spoilage). The refrigerated transport of perishable products provides a critical link in the cold chain between the consumer and the producers, processors, shippers, distributors and retailers. The transport refrigeration industry incorporated mechanical vapour compression mechanisms into the first transport refrigeration units. Therefore, the physical, chemical and thermodynamic properties of these refrigerants had a major impact on the design of compressors and other refrigeration system components. Many refrigerants have excellent thermodynamic properties suitable for use in vapour compression refrigeration systems, but are limited by other properties such as toxicity, flammability and chemical stability within the refrigeration system. Most district heating systems in former planned economies are less efficient and oversized. In other words, their supply infrastructure is larger than necessary to meet current demand. This problem can be exacerbated when they lose customers. Reforms, particularly for tariffs, are needed to improve finances and break the vicious circle of deteriorating competitiveness.

Chapter 9

CONCLUSION

We love in a society of unprecedented: consumption is the norm. Nevertheless, people are only just starting to be concerned about the process that gets products onto the shelves and the effect of their use on our planet. The massive increases in fuel prices over the last years have however, made any scheme not requiring fuel appear to be more attractive and to be worth reinvestigation. Economic projections are difficult at the best of times, when economies are relatively stable and a reference 'business as usual' case can be used. However, there are numerous signals that the world faces very turbulent economic conditions for a while -a credit crunch may make some project finance difficult and the shortage of raw materials could lead to supply chain difficulties. However, the rapidly escalating price of oil is focusing a lot of attention on the price of energy and the hedge of electricity supply without a fuel cost is likely to become increasingly attractive to many companies and utilities. At some stage, rising fuel costs could lead to demand for wind energy becoming almost infinite. The main factors expected to influence the continuing growth of the energy sector are:

The economies of the transition states (Russia and Central Asia) will start to grow.

Increasing energy demand in Asia and South America.

Oil prices will continue to remain high as will demand for fossil fuels.

Continuing competitiveness of renewables with fossil fuels.

Many countries may find they are well off their international CO_2 reduction commitments and need to install some new renewable capacity very quickly.

Security of supply questions will continue to support renewable technologies.

Deregulated markets will remove excess conventional power capacity and new capacity is likely to be more expensive than wind.

Newspapers, TV, schools, universities and politicians rant and rave about being ‘green’ and doing our bit for the environment, but can we as individuals change things? Energy efficiency brings health, productivity, safety, comfort and savings to homeowner, as well as local and global environmental benefits. The use of renewable energy resources could play an important role in this context, especially with regard to responsible and sustainable development. It represents an excellent opportunity to offer a higher standard of living to local people and will save local and regional resources. Implementation of greenhouses offers a chance for maintenance and repair services. It is expected that the pace of implementation will increase and the quality of work to improve in addition to building the capacity of the private and district staff in contracting procedures. The financial accountability is important and more transparent. Various passive techniques have been put in perspective, and energy saving passive strategies can be seen to reduce interior temperature increase thermal comfort, and reducing air conditioning loads. The scheme can also be employed to analyse the marginal contribution of each specific passive measure working under realistic conditions in combination with the other housing elements. In regions where heating is important during winter months, the use of top-light solar passive strategies for spaces without an equator-facing façade can efficiently reduce energy consumption for heating, lighting and ventilation.

RECOMMENDATIONS

The current analysis does support the below-mentioned recommendations, which are also drawn from the author own experience. However, further research is needed to find out if such recommendations might be effective:

Launching of public awareness campaigns among local investors particularly small-scale entrepreneurs and end users of renewable energy technology (RET) to highlight the importance and benefits of renewable, particularly solar and wind energy resources.

Amendment of the encouragement of investment act, to include furthers concessions, facilities, tax holidays and preferential treatment to attract national and foreign capital investment.

Allocation of a specific percentage of soft loans and grants obtained by governments to augment budgets of research and development (R and D) related to manufacturing and commercialisation of RET.

Governments should give incentives to encourage the household sector to use solar and wind energy instead of conventional energy.

Execute joint investments between the private-sector and the financing entities to disseminate the renewable with technical support from the research and development entities.

Availing of training opportunities to personnel at different levels in donor countries and other developing countries to make use of their wide experience in application and commercialisation of RET particularly solar and wind energy devices.

The governments should play a leading role in adopting solar and wind energy devices in public institutions, e.g., schools, hospitals, government departments, police stations etc., for lighting, water pumping, water heating, communication and refrigeration.

To encourage the private-sector to assemble, installs, repair and manufacture solar and wind energy devices via investment encouragement, more flexible licensing procedures.

REFERENCES

[1] Advanced Energy. Information available at: /http://www. advanced energy. org/motors_and_drives/knowledge_library/resources/permanent_magnet_m otors.htmlS

[2] High Technology Finland. Information Available at: / http:// 2003. high techfinland.com/2003/2003/energy environment/energy/abbell.htmlS

[3] Muller S, Deicke M, De Doncker RW. Doubly fed induction generator systems for wind turbines. *IEEE Industry applications magazine*, 2002; 8 (3): 26–33.

[4] Grabic S, Katic V. A comparison and trade-offs between induction generator control options for variable speed wind turbine applications. In: *Proceedings of IEEE international conference on industrial technology*, vol. 1, December 2004. p. 564–8.

[5] Chondrogiannis S, Barnes M, Aten M. Technologies for integrating wind farms to the grid. New and Renewable Energy Programme, *Department of Trade and Industry*, February 2006.

[6] Meier S, Norrga S, Nee HP. New topology for more efficient offshore AC/DC converters for future offshore wind farms. *KTH Research Project Database*, February 2006.

[7] Henderson G, Roding W. Synchronous and synchronised wind power generation. In: Proceedings of eighths NZWEA conference, Palmerston North, July 2004.

[8] Barry D. Increasing renewable energy accessibility in Ireland. In: 20th wind energy congress, September 2006.

[9] Jamal A., et al. A review of power converter topologies for wind generators. *Renewable Energy*, 32 (14): 2369-2385.

[10] Omer, A.M. Wind speeds and wind power potential in Sudan. In: Proceedings of the 4th Arab International Solar Energy Conference. Amman: Jordan. 1993.

[11] World Meteorological Organisation (WMO). Guide to meteorological instrument and observing practices. 4th Edition. WMO. No.8, TP.3, Geneva: Switzerland. 1994.

[12] Eldridge, F.R. Wind machines. 2nd Edition. Van Nostrand Reinhold. New York: USA. 1980.

[13] Smedman-Högström, A., and Högström, U. Practical methods of determining wind frequency distribution for lowest 200 meters from routine meteorological data. *Journal of Applied Meteorology*, Vol.7. 1978.

[14] ESDU 72026. Characteristics of wind speed in the lowest layers of the atmosphere near the ground: strong winds. Eng. Sci. Data Unit Ltd, 251 Regent Street, London WIR 7AD: UK. 1972.

[15] Monin, A.S., and Obukov, M.A. Dimensionless characteristics of turbulence in the surface layer. Akad. Nank. SSSR. Geofiz Inst. 1954.

[16] Wong, R.K. Weibull distribution, iterative likelihood techniques and hydrometerological data. Journal of Applied Meteorology. Vol.16, p.1360-1364. 1977.

[17] Pasquill, F. Some aspects of boundary layer description. *Quarterly Journal of the Royal Meteorological Society*. Vol.18. 1972.

[18] Lysen, E.H. Introduction to wind energy. The Netherlands: *CWD*. p.15-50. 1983.

[19] Stevens, M.J.M., and Smulders, P.T. The estimation of the parameters of the Weibull wind speed distribution for wind energy utilisation purposes. Wind Energy Group. Department of Physics. University of Technology. Eindhoven: The Netherlands. *Wind Engineering*, 3(2): 132-145. 1979.

[20] Justus, C.G. Winds and wind system performance. The Franklin Institute Press. 1978.

[21] Duchin, F. Global scenarios about lifestyle and technology, the sustainable future of the global system. United Nations University. Tokyo. 1995.

[22] World Energy Outlook. International Energy Agency. OECD Publications. 2 rue André Pascal. Paris. France. 1995.

[23] VAN Schijndel, P., Den Boer, J., Janssen, F., Mrema, G., and Mwaba, M. Exergy analysis as a tool for energy efficiency improvements in Tanzania and Zambia industries. ICESD Conference Engineering for Sustainable development. University of Dar Es Salaam. Tanzania. July 27-29, 1998.

[24] IPCC. Climate change 2001 (3 volumes). United Nations International Panel on Climate Change. Cambridge University Press. UK. 2001.

[25] UNIDO. Changing courses sustainable industrial development, as a response to agenda 21. Vienna. 1997.

[26] DEFRA, Energy Resources. *Sustainable Development and Environment.* UK. 2002.

[27] Parikn, J., Smith, K., and Laxmi, V. Indoor air pollution: a reflection on gender bias. *Economic and Political Weekly*. 1999.

[28] Steele J. Sustainable architecture: principles, paradigms, and case studies. New York: McGraw-Hill Inc., 1997.

[29] Sitarz D. Editor. Agenda 21: The Earth Summit Strategy to save our planet. Boulder (CO): Earth Press. 1992.

[30] Rodda, J.C. Water under pressure. Hydrological Sciences- *Journal des Sciences Hydrologiques,* 46 (6): 841-854. 2001.

APPENDIX

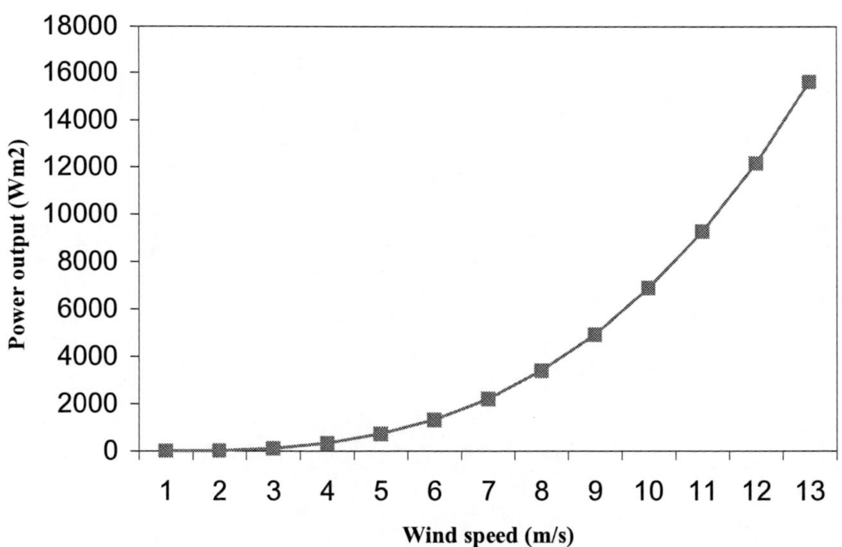

Wind speed vs Power output
18000
16000
14000
12000
10000
8000
6000
4000
2000
0
Power output (Wm2)
1 2 3 4 5 6 7 8 9 10 11 12 13
Wind speed (m/s)

INDEX

A

B

C

D

E

F

G

H

I

J

L

M

N

O

T

U

V

W

Y